FROGS

SEYMOUR SIMON

HARPER
An Imprint of HarperCollinsPublishers

This book is dedicated to the special bonding that takes place between kids and their caregivers who explore together the wonders of frogs in the spring.

PHOTO CREDITS

Page 2: © Naturfoto Honal/Corbis; page 4: © Eric Isselee; page 6: © Brian Lasenby; page 7: © Animals Animals/SuperStock; page 8: © Valerii Kirsanov; page 9: © Galyna Andrushko; page 10: © Hunsa Prachansuwan; page 11: © Heidi & Hans-Juergen Koch/Minden Pictures/Corbis; page 12: © Carmina Mcconnell; page 14: © tonarinokeroro; page 15: © Nature PL/SuperStock; page 17: © hotshotsworldwide; page 18: © imagebroker.net/SuperStock; page 19: © Cathy Keifer; page 20: © Chris Hill; page 21: © Flirt/SuperStock; page 22: © Stephen Dalton; page 23: © Oleg Tovkach; pages 24–25: © NHPA/SuperStock; pages 26–27: © Benoit Daoust; page 28: © Marie Daloia; page 30, top to bottom: © Cathy Keifer; © Brian Lasenby; page 31: © donyanedomam; pages 32–33: © Minden Pictures/SuperStock; page 34, top to bottom: © Mrs. Ya; © age footstock/SuperStock; page 35: © hotshotsworldwide; page 36: © STR/Reuters/Corbis; pages 38–39: © hotshotsworldwide

Frogs

ISBN 978-0-06-228912-4 (trade bdg.) — ISBN 978-0-06-228911-7 (pbk.)

15 16 17 18 19   SCP   10 9 8 7 6 5 4 3 2 1

❖

First Edition

Author's Note

From a young age, I was interested in animals, space, my surroundings—all the natural sciences. When I was a teenager, I became the president of a nationwide junior astronomy club with a thousand members. After college, I became a classroom teacher for nearly twenty-five years while also writing articles and books for children on science and nature even before I became a full-time writer. My experience as a teacher gives me the ability to understand how to reach my young readers and get them interested in the world around us.

I've written more than 250 books, and I've thought a lot about different ways to encourage interest in the natural world, as well as how to show the joys of nonfiction. When I write, I use comparisons to help explain unfamiliar ideas, complex concepts, and impossibly large numbers. I try to engage your senses and imagination to set the scene and to make science fun. For example, in *Penguins*, I emphasize the playful nature of these creatures on the very first page by mentioning how penguins excel at swimming and diving. I use strong verbs to enhance understanding. I make use of descriptive detail and ask questions that anticipate what you may be thinking (sometimes right at the start of the book).

Many of my books are photo-essays, which use extraordinary photographs to amplify and expand the text, creating different and engaging ways of exploring nonfiction. You'll also find a glossary, an index, and website and research recommendations in most of my books, which make them ideal for enhancing your reading and learning experience. As William Blake wrote in his poem, I want my readers "to see a world in a grain of sand, / And a heaven in a wild flower, / Hold infinity in the palm of your hand, / And eternity in an hour."

Seymour Simon

Frogs are **amphibians**, animals that live in water and on land. Even when living on land, frogs don't drink water. Instead, they soak water into their bodies through their moist skins. Frogs look like frogs only part of their lives. For the first part of their lives, they don't have arms or legs and they look more like fish. Newborn frogs, called **tadpoles**, live underwater and even have **gills** like fish. But something strange and wonderful happens as they become adults. They lose their gills, grow arms and legs, and come on land as frogs.

Early spring is the best time to look for frogs in **temperate** regions. In the central and eastern parts of the United States and Canada, the sound of frogs is the signal that spring is coming. Just after the snow melts, you can hear the sounds of frogs called spring peepers. Spring peepers are small frogs that are less than an inch and a half in length.

Peepers wake up from their winter sleep and begin to peep loudly from early March to May or June. They have **sacs** under their mouths that inflate like little balloons. When the air is released from the sac, it makes a sound like a high-pitched whistle, but a chorus of peeps is like an orchestra of jingling bells. The chorus is the sound of hundreds of male spring peepers calling for mates. If you go close to a pond full of peepers, they fall silent and they're hard to see. Spring peepers begin to sing at sundown because they are most active at night.

Frogs are **exothermic**, or cold-blooded, which means that their body temperature changes with their surroundings. They are cold when the water or air around them is cold, and they are warm when the water or air around them is warm.

Just like their ancestors, frogs are **adaptable** to different conditions and have been on Earth for over 140 million years. Scientists know this because they have found frog **fossils** that date back to the Jurassic period, when dinosaurs roamed the land.

Different kinds of frogs live all over the world and on every continent except ice-covered Antarctica. They usually live near slow-moving bodies of water, or **wetlands**, such as lakes, ponds, and marshes. But some kinds of frogs live in fast-moving waters, such as waterfalls, while other kinds live in dry deserts. In deserts, frogs burrow into the ground and stay in a kind of sleep, called **estivation**, until it rains. Frogs do not live in the oceans or other bodies of salt water.

Frogs in temperate areas come out of their winter sleep in the early spring when the water temperature goes over 41° Fahrenheit. Male frogs begin to **croak** in the spring in the warming waters of a pond. They use their front legs to clasp female frogs that were attracted by the croaking. The eggs are laid by the female frog directly into the water and are then fertilized by sperm cells from the male.

Temperate-region frogs lay thousands of eggs. Each egg is about the size of a pinhead and surrounded by a jelly-like substance. The jelly swells up in the water and sticks together with other eggs in large clumps. The eggs take a few days to three weeks to hatch. Most kinds of frogs abandon their eggs once they are in water, but not all eggs hatch by themselves. Marsupial frogs carry their eggs with them. Glass frogs guard their eggs by sitting near or on top of them.

Looking at frog eggs with a magnifying lens opens up new visions. The egg is black above and white below. The white is the egg yolk that supplies food to the dark developing **embryo**. The egg floats dark-side up and is warmed by sunlight, while the white yolk faces downward, making the egg more difficult to see.

Newly hatched tadpoles don't look like frogs at all. They are tiny animals with no mouths, nostrils, or legs. Like fish, they have outside gills, which let them get oxygen directly from the water. A young tadpole spends most of its time attached to plants, egg jelly, or rocks in the water.

For the first few days, tadpoles feed on their own egg yolks. During that time, their tails grow longer and their outside gills form. They begin to swim freely and feed on water plants. Tadpoles are a favorite food of fish, water beetles, and many other animals that hunt them in ponds and lakes. Tadpoles are fast swimmers, but only a few will escape their **predators** and grow into adults.

During the next few weeks of a tadpole's life, a great change, called **metamorphosis**, takes place. First, small back legs appear and begin to grow larger. Toes form, and the legs thicken and begin to look like frogs' legs. The tadpole begins to use its legs to swim. After the hind legs grow, tiny front legs appear. As soon as the front legs appear, the tail begins to shrink. The eyes bulge out and the tadpole starts to look like a frog.

Inside the tadpole, the gills disappear and lungs develop. The tadpole surfaces every so often to take a breath of air. After a while the tail is completely gone and the tadpole is no more. In its place is a complete tiny **froglet**. The froglet needs to come out on a rock or a bit of land so that it can rest while breathing air or it might drown.

In nature, most kinds of frogs live along the banks of ponds, lakes, and streams. Others live in fields, woodlands, and wetlands. Still others live up in trees, hidden in underground burrows, or in places distant from water. Wherever frogs live as adults, almost all come back to the water to mate and lay their eggs to begin the cycle of metamorphosis once again.

Like most animals, frogs have five main senses: sight, sound, touch, taste, and smell. Frogs have good eyesight. They will snap at any small moving object they see. Their eyes bulge out from their heads so they can see forward and backward and sideways all at the same time. When a frog floats just below the surface of the water in a pond, its eyes are like the **periscope** of a submarine. It's on the lookout above the water for food or enemies. The frog is well **camouflaged** in this position and can barely be seen from above or below.

The size and movement of an object is important to a frog. Frogs do not snap at objects that are too large or are not in motion. For example, frogs snap at insects that buzz near them. But if a large object such as yourself moves near a frog sitting on the banks of a pond, it will stop croaking and leap into the water. If you stay still or move very slowly, you may be able to get close up to a frog without disturbing it.

Frogs even use their eyes to help swallow food. When a frog blinks, its eyeballs push downward, making a bulge in the top of its mouth. The bulge squeezes the food down to the back of its throat.

A frog's eyes are covered by lower transparent eyelids. They protect the eyes underwater and keep them moist on dry land. If something touches the eyelid, the frog responds by drawing the eyelid up and pulling the eye back into its head.

Frogs have good hearing, but they don't have ears that stick out on the outside of their bodies the way you do. Their eardrums are level with the frog's skin and are on either side of the head. You can tell the difference between a male and a female frog by the size of their eardrums. In males, eardrums are about the size of the eyes. In females, eardrums are smaller than the eyes. Hearing is important to frogs. Male frogs attract female frogs with a croaking song.

Frogs also have a sense of touch. Stroke a frog's skin very gently with the tip of a blade of grass and it may not move. But poke it with a finger and the frog will jump away. Frogs have a touch-sensitive lateral line along their bodies. When frogs swim in a river, they pick up

the pressure of the rushing water on their lateral lines and respond by swimming against the current. Frogs can also sense temperature changes and dry conditions through their skins. If a frog

sits in the sun for too long, it just hops away into the shade or takes a dip in the cool water.

Frogs have other senses as well. When a frog puts bitter insects in its mouth, such as acid-producing red ants, it will spit them out quickly. Frogs use their sticky tongues to catch and swallow insects. Unlike yours, a frog's tongue is not attached to the back of its mouth. A frog's tongue is attached to the front of its mouth, which allows a frog's tongue to reach out much farther than your tongue.

Think about how you would describe one thing a frog does and you might say, "Frogs jump." A frog's hind legs are its power legs. When a frog is at rest, it folds its hind legs underneath and against its body. When it is going to jump, the muscles tense and the frog pushes its hind legs against the ground and launches into the air. The frog travels several feet in the air and lands first on its small front legs.

How far can a frog jump? Most kinds of frogs can jump far, but some are world-champion jumpers. Leopard frogs can jump at least 30 times their body length, which means that they can make jumps of over ten feet. Small tree frogs can jump 50 times their body length. That is like a person jumping from one end of a football field to the other end in a single bound! Flying frogs of Asia have webbed feet that they use as parachutes to glide from one tree to the next. Jumping is a good way of catching **prey** and can help frogs escape from an enemy.

Frogs swim using their powerful hind legs. Naturally, they do the "frog kick" when swimming. They fold their hind legs close to their bodies, then stretch out quickly. Five long toes on each foot are connected by webbing. Their webbed feet act like paddles pushing against the water. It's no wonder we call military-trained underwater divers frogmen.

By early autumn, leopard frogs and bullfrogs appear around the edges of ponds and streams. Temperatures soon start to drop as the days grow shorter. Winter is coming. Before the ponds freeze over, the frogs go into the water and begin their winter sleep, called **hibernation**. Hibernation is when an animal slows down its breathing and body movement so that it can live through the cold winter by using the energy stored in its body. During hibernation, a frog's eyes are covered by the transparent lower lid. It doesn't respond easily to touch. In temperate regions, most kinds of frogs hibernate.

Hibernating frogs spend their winter sleep lying on top of the mud at the bottom of the pond. They get oxygen from the water and don't need to come up for air. At times, they slowly swim around. Frogs would starve or freeze to death if they did not hibernate in winter. When spring comes, temperatures rise and the ice melts. Frogs now wake up and go on with their lives.

Frogs that spend their lives on land, such as the wood frog and the spring peeper, hibernate in deep cracks in logs or rocks or deep in leaf litter. These frogs are not as well protected from the cold as underwater frogs and may freeze. They can still survive because they have a high concentration of **glucose**, a kind of sugary antifreeze, in their blood.

There are three different groups of amphibians in the world: frogs and toads, salamanders and newts, and a rare group called caecilian (see-**sil**-ee-uhn) (legless and sightless animals that look like worms). Frogs and toads are by far the most common amphibians and the ones that you see most often. Frogs and toads look very much alike, so deciding which is which is not always easy.

Most frogs have bulging eyes, smooth skin, long hind legs, and webbed feet and live in or near water. Most toads live on land, have dry skin with warts and stubby bodies with short hind legs, and are less active. But some frogs don't live near water and have no webbing, and some toads have smooth skin. So the word "frog" is sometimes used for both frogs and toads.

Most people say "frogs" when they are talking about members of a family called *Ranidae*, which includes leopard frogs, bullfrogs, green frogs, pickerel frogs, and wood frogs. They are sometimes called "true frogs" because their body forms and their life stories are so similar. "True toads" are members of the family called *Bufonidae*, and they have poison glands behind their eyes. The poison is a milky white toxin that toads squeeze onto the surface of their skin when threatened. When swallowed, the poison is dangerous to dogs, cats, and even humans.

Here are some common frogs found in temperate regions:

- Leopard frogs are mostly green and brown with a white underside. They are named "leopard" because of the dark spots across their backs. They were once the most abundant frog in North America, but they are far fewer in number since the 1970s, probably because of environmental  reasons such as acid rain and destruction of **habitats**.

- Bullfrogs are the largest frogs in North America. The females are  larger than the males and grow up to 8 inches in length. They spend their lives in and around water. They are green above and white or cream colored underneath. Bullfrogs live alone and are very territorial. They rarely meet except to fight or mate.

The females lay as many as 20,000 eggs in huge sheets attached to underwater plants.

- Wood frogs have a black band that stretches past both eyes to the eardrums and looks like a bandit's mask. They are one of the first kinds of frogs to wake up and breed in early spring, sometimes when the pond is still partly frozen. The frogs sound like quacking ducks and can be heard from far away. After a few days of breeding, the eggs are left in masses in the pond and the adult frogs go off into the woods. They will return to lay eggs again the following spring.

Here are some unusual frogs from around the world:

- Darwin's frogs live in the cool forest streams of South America. They are small and green and don't look unusual, but they have very odd breeding behavior. The female lays about three dozen eggs and the male guards them for a few weeks until they are ready to hatch. Then the male Darwin's frog carries around the developing tadpoles in its vocal pouch. When the tadpoles develop into tiny frogs, they are spat out in water and they swim away.

- Poison dart frogs are a family of some of the most beautiful colored frogs in the world. Depending on the particular species, these frogs can be red, blue, green, yellow, gold, or a mixture. Their exotic colors and patterns warn other animals not to eat them because they  are poisonous. The males are great parents, carrying both eggs and tadpoles on their backs.

- Amazon horned frogs are about the size of a grapefruit, much bigger than the frogs in a pond near you. These big and round frogs are found in marshes and swamps in the rainforests of the  Amazon basin. They hide in the **leaf litter** in the forest so that only their heads stick out. They will eat any small animal and even try to eat animals that are too big to be swallowed whole. That's why they are sometimes called Pac Man frogs.

Here are some interesting toads from around the world:

- American toads are the most common toads in the United States. They are usually brown, olive, or brick red and have warts all over their bodies. American toads live in forests, fields, lawns, and gardens. They hop across  roads in the spring breeding season and in the fall when they are looking for places to hibernate. The toad call is a long trill that sounds much like a cricket's song. You can tell which is which because toads sing in spring and crickets chirp in fall. American toads eat insects, earthworms, spiders, and just about any small creature that wanders by.

- Midwife toad males carry a string of eggs wrapped around their ankles. The eggs develop for several weeks before they are released into ponds as tadpoles. These toads live in

Germany, France, and other mid-European countries.

• Fire-bellied toads don't look like much from above—an average-sized green toad with black spots. But when it feels threatened, the fire-bellied toad rises up and arches its back to display a brilliant red-and-black belly. This display warns predators that the toad is poisonous and that they had better not eat it. This toad is common in streams and ponds in parts of China, Korea, Japan, and Russia.

Climate change and global warming may have affected frogs' ability to keep their bodies cool and wet enough. Because frogs absorb water through their skin, they are at risk of dying from water pollution and acid rain. Chemicals, fertilizer, sewage runoff, and other kinds of man-made pollution threaten frogs.

The number of frogs is declining in many places around the world. A major worldwide threat to frogs today is the chytrid fungus. The fungus feeds on keratin, a substance in frogs' skin that makes it tough. The fungus does not usually affect tadpoles, but it can kill adult frogs. Scientists think that about one-third of the world's frog population may have the fungus. There is no effective treatment, so some scientists are trying to **quarantine** and keep safe as many frog species as possible in zoos and other facilities.

Chytrid fungus is a worldwide problem, but there are a few things you can do to help stop the spread. You should never release foreign pet frogs into the wild. Released foreign frogs can breed and overwhelm the local frog population. They can also spread disease. If you spot many sick or dead frogs in your area, contact your state or the federal wildlife department and tell them about what you've found. The information may be helpful to scientists studying the problem.

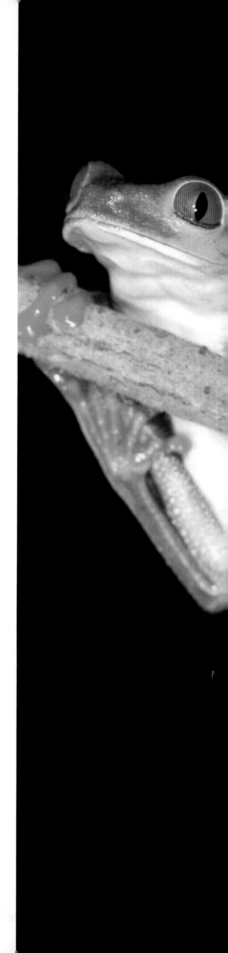

Frogs play an important role in the balance of nature. They help to control the insect population. They are an important food source for many other animals, such as birds and snakes. And, because of their double life as tadpoles in water and adults on land, even common frogs are among the most interesting animals in the world.

GLOSSARY

Adaptable—Able to adjust to different conditions.

Amphibians—Animals that live in water and on land.

Camouflage—Changing one's appearance to blend into one's surroundings.

Croak—A low-pitched sound.

Embryo—The very early developmental stages of birth or hatching.

Estivation—When an animal's body slows down and goes into a deep sleep during the hot, dry months of summer.

Exothermic—Cold-blooded.

Fossil—The remains or impression of a once living thing.

Froglet—The stage between tadpole and adult frog.

Gills—A respiratory organ so that aquatic animals can breathe underwater.

Glucose—A kind of sugary antifreeze in blood.

Habitat—The natural environment of a person, animal, or organism.

Hibernation—When an animal slows down its breathing and body movement so that it can live through the cold winter by using the energy stored in its body.

Leaf litter—Dead plants, such as leaves, that have fallen to the ground.

Metamorphosis—A great change in form that takes place.

Periscope—An instrument consisting of a tube with prisms that is used to view object that are above the direct line of sight or obstructed from view.

Predator—An animal that kills or preys on other animals.

Prey—An animal that's hunted or captured for food.

Quarantine—Isolation to prevent the sprea of a disease.

Sac—A baglike structure that contains flui

Tadpoles—Newborn frogs.

Temperate—Not extreme weather in regar to hot or cold.

Wetlands—Slow-moving bodies of water, such as lakes, ponds, and marshes.

INDEX

Bold type indicates illustrations.